VHF RADIOTELEPHONY
FOR
YACHTSMEN

Published by the ROYAL YACHTING ASSOCIATION, RYA House, Romsey Road, Eastleigh, Hampshire SO50 9YA. Tel: 01703 627400 and printed by Blackmore, Shaftesbury, Dorset

CONTENTS

INTRODUCTION

Radiotelephones in small sailing yachts are now becoming much more usual and as a result of the increase in the number of operators the discipline of radiotelephony procedure is becoming more and more important.

In many respects it is a pity that the word "telephone" was ever used because VHF is really a massive "party line". Sometimes a radiotelephone *is* used as a telephone since it enables a yachtsman at sea to talk to his home or office by being "plugged in" to the shore telephone system. But, at sea, VHF R/T is more — it is a complete communication system but it differs from other forms of radio communication in two major aspects. Firstly, it is very seriously restricted in range, although, for some purposes, this restriction is a positive advantage because the same frequency can be shared by dozens of stations as long as they are not too close to each other. The second major difference between VHF R/T and other types of radio communication is what is known as the "capture" effect. If two stations are transmitting simultaneously and a third is within range of both, one of the transmitters will dominate the conversation and completely exclude the other. Thus one conversation may be completely wiped out by another transmission cutting in. The existence of this "capture effect" makes it essential that proper discipline be observed.

There are rules laid down for the procedure and, as in every other sphere of life, there are operating procedures which "grow up" and eventually become the accepted "norm" although they may not be entirely in accordance with rules which, in some cases, were laid down some years ago.
We have, in this book, attempted clearly to distinguish between rules and operating procedures which are acceptable to the authorities and we have set out what is today's common practice.
There is, of course, a world of difference between the highly professional activities of the ship's radio operator and the requirements imposed upon him by the rules and the activities of the average practising yacht owner equipped

with a VHF radiotelephone. However it behoves the "amateur" to be as "professional" as possible and we hope that the contents of this booklet will assist all those people who make use of this excellent modern aid.

The correct procedure is well worth learning and well worth teaching to members of your crew. Five minutes spent so that everyone on board your yacht will know how to operate your radiotelephone will be five minutes well spent.

Thanks are due to many people for their work in preparing the text of this booklet but particularly to Peter Fulton, who had the "original" idea of having a teaching programme running throughout the text of the book. Various committee members of the RYA have given practical advice on the use of VHF in yachts and members of the staff of the Home Office, British Telecom and the Post Office have all been most cooperative in advising those responsible for the production of this publication.

DISTRESS

The definition of distress was amended, by the 1979 Search and Rescue Convention, to include danger to life. The definition is now:

"Grave and imminent danger to a ship, aircraft, vehicle or person, requiring immediate assistance".

CHAPTER I

LICENSING REQUIREMENTS
AND
OPERATING REGULATIONS

Applications for Licences

Applications for any of the Ship licences (but not the Operator's Certificate of Competence) should be made to the Department of Trade and Industry, Licensing Branch, Radio Regulatory Division, Waterloo Bridge House, London SE1 8UA. Applicants will be notified of the amount of the licence fee currently payable. The installation or use of any radio transmitting apparatus except in accordance with the terms of a licence is an offence under the Wireless Telegraphy Act.

Transmission and Reception Licences

The use of a transmitter on board a yacht is subject to three distinct licensing requirements, for the set, the installation and the operator.

Technical Requirements for Equipment

The Department of Trade and Industry lays down Type Approval Standards for radiotelephone equipment. These standards are designed to ensure that equipment cannot inadvertently cause interference with other radio users. All radiotelephones sold by reputable suppliers meet the Department of Trade and Industry Type Approval standards. There have, however, been instances of second-hand and military surplus sets which do not meet Type Approval standards being offered for sale. If you decide to buy one of these sets it is important to check, with the Department of Trade and Industry Radio Regulatory Department, that the model has been tested for Type Approval.

The Ship (VHF) Licence

A 'Ship Licence' authorises the use of a transmitting and receiving radio in a yacht. This licence authorises two-way communication with other vessels and Coast Radio Stations on frequencies in the international maritime band. Specific frequencies are allocated for six major purposes as described in Chapter II.

There are a number of conditions attached to the issue of the licence, one of which is that the radio must be operated only under the control of an operator holding an appropriate certificate of competence. Other conditions, relating to frequencies which may be used and maximum power output reflect the Department of Trade and Industry's responsibilities under international agreement to ensure the orderly and economical use of radio and the avoidance of interference.

Yachts which carry MF or HF radiotelephones require a full Ship Licence, which authorises the use of these frequencies.

The Ship (Emergency Only) Licence
If the only requirement for two-way communications is for emergency use, a Ship (Emergency Only) Licence may be obtained. This licence is granted in the same way as the full 'Ship Licence' but it authorises communication only to call for assistance in an emergency, to alert the search and rescue organisations and to communicate with ships and aircraft answering a distress call.
Portable equipment is available (at relatively modest cost) which operates on 2182 kHz. A licence is necessary and is valid for five years.

Emergency Position Indicating Radio Beacons
"EPIRBS", as they are phonetically spoken of, are automatic beacons transmitting on 'aeronautical' or marine band frequencies and also need a licence.

The Operator's Certificate of Competence
A two-way radio installation in any vessel must be operated only under the control of the holder of an appropriate certificate of competence and authority to operate.
In a yacht with a VHF radiotelephone only, the appropriate certificate is a 'Restricted Certificate of Competence in Radiotelephony (VHF only)'. This certificate states that the holder has passed the examination in Radiotelephony required under the provisions of Section 7(1) of the Wireless Telegraphy Act 1949 and the Radio Regulations and that the holder has made a declaration that he will preserve the secrecy of correspondence.
The examination for the award of the certificate consists of

7

a written test followed by a practical test of the candidates's knowledge of VHF radiotelephone procedures. Candidates are required to have a thorough knowledge of distress, urgency and safety procedures and a working knowledge of all other operating procedures and the regulations which apply to Marine Band VHF radiotelephony.

The syllabus, full details of the form of examination and addresses and telephone numbers of examination centres are contained in RYA booklet G26, VHF Radio Operator Examinations. Applicants for this examination should approach the nearest examination centre. A list of these is contained in Appendix C.

Channel M (A private frequency)
The Department of Trade and Industry may grant an authorisation to the holder of a Ship Licence to use Channel M, (157.850 MHz) a frequency outside the international maritime band, for communication with U.K. yacht clubs or marinas on matters of club or marina business. There is no fee payable for this authorisation if a full radio licence is already held. Nor is there any additional requirement for operator qualification and no fee is payable for this authorisation.

Otherwise you need to obtain a Private Mobile Radio Licence which entitles you to operate one base station and up to nine mobile stations on Channel M, provided the operators hold a Restricted Certificate of Competence.

Channel M2(161.425MHz)
This private channel has been allocated as the primary working channel for Yacht Clubs. In cases of overload or non-availability Channel M may be used. Ship Licence holders authorised to operate on Channel M may operate on Channel M2 without further authorisation.

The Radiotelephone Log
Under the Radio Regulations, every ship fitted with a radiotelephone installation should carry and keep an up-to-date radiotelephone log (Diary of Radio Service).

The log should provide a record of all messages sent and received, including details of the time (GMT), sender, recipient, channel used, and a brief record of the content of the message. Entries in respect of messages or radiotelephone calls in the public correspondence service should be restricted to the call, serial number of the message or call and time received or sent.

Use of the Radiotelephone When in Port
Strict rules are necessary to control the use of radiotelephones in port. Without these rules the high shipping density in and around the approaches to a busy harbour would result in overloading the available frequencies to the extent that it would become impossible to pass essential messages.

In UK harbours and estuaries a radiotelephone may be used only:

i) For Port Operations Service communications.

ii) On private channels licensed by the Department of Trade and Industry for a specific purpose (i.e. channel M).

iii) For the exchange of traffic through the nearest Coast Radio Station.

Intership communication is permitted only on matters relating to safety.

Marina Communication
Channel 80 has been allocated as the primary working channel for Marinas in the United Kingdom. In cases of overload or non-availability Channel M may be used.

Port Operations
Communications on port operations channels must be restricted to those relating to operational handling, the movement and the safety of ships and, in emergency, to the safety of persons. Public correspondence messages must not be passed on channels designated for port operations or ship movement.

The ability to communicate direct with harbour authorities is particularly useful in busy commercial ports. At Dover, for instance, every vessel wishing to enter or leave harbour must first obtain permission to do so. In a yacht not fitted with VHF this entails an exchange of signals in morse by light, or by flags, neither an easy procedure from a yacht in heavy seas. With a VHF radio, however, obtaining permission to enter or leave is much simpler and by monitoring the port operations channel the yachtsman will be able to build up a picture of the traffic movements in and out of the harbour.

Documents to be Carried

Ship stations voluntarily fitted with a radiotelephone installation are required to carry the following documents:

The Ship Licence.

The Certificate(s) of the Operator.

This will normally be the Restricted VHF Only Radio Operator's Certificate of Competence.

A list of Coast Radio Stations with which communications are likely to be conducted. Available from Yachtsmens' Nautical Almanacs or the Admiralty List of Radio Signals Vol. 1.

The Radiotelephone Log

In Appendix D you will find the detail contained in all the publications which may be useful to yachtsmen with a VHF Radiotelephone on board.

CHAPTER II

THE INTERNATIONAL ORGANISATION
OF MARITIME RADIOTELEPHONE
COMMUNICATIONS

The VHF International Maritime Mobile Band

The VHF frequencies between 156.00 MHz and 174.00 MHz are allocated by international agreement for the Maritime Mobile Service, in other words for use by ships fitted with VHF radio. This allocation is not exclusive, and other services are included in the band.

The part of this band allocated to the Maritime Mobile Band was subdivided into 28 channels some years ago. Additional channels introduced in 1972 were interleaved between existing channels, with the channels between 29 and 59 being used for a number of other services. Hence the apparently illogical numbering of channels.

A full list of the channels will be found in Appendix A. Many of the channels have two frequencies allocated, to allow 'duplex' working. This term (and its implications) are explained later.

Each channel is allocated for one or more of six specific purposes, distress safety and calling; intership; public correspondence; port operations; ship movement (very similar and for all practical purposes identical to port operations); and in the U.K. only, yacht safety. This allocation is made by international agreement in order to introduce order into what would otherwise be a chaotic situation.

Distress Safety and Calling Frequency (Channel 16)

All ships are encouraged to maintain a continuous watch on Channel 16 when at sea because it is the VHF Distress Safety and Calling frequency. In many cases it is the Channel upon which you establish initial contact and then, by arrangement, move to a working frequency. During the last few years the number of radiotelephones in use at sea has increased to the extent that in the busiest areas there is great pressure on Channel 16. To relieve this pressure there is a growing tendency to use working frequencies for initial calls but this can, of course, only be done if the station called is maintaining a listening watch on a working frequency.

All Coast Radio Stations expect initial calls on working frequencies. Most port and harbour authorities maintain continuous watch on their working frequencies and can be called direct. The Coastguard, on the other hand, maintain continuous VHF watch only on Channel 16, so initial calls should be on that frequency.

Many VHF transceivers are now fitted with dual watch, which enables Channel 16 and another frequency, to be monitored. With dual watch it is possible to listen to Channel 16 and, say, an intership frequency. If you are expecting a call from another yacht it is sensible to have an arrangement to listen on an intership frequency so that contact can be established direct without the need to use Channel 16.

Port Operations Service

A number of international VHF channels are designated exclusively for port operations use and several others are available on a shared basis. Channels 12 and 14 are the two most commonly used. Channels are allocated in such a way as to minimise mutual interference between harbours which are within VHF range of each other. Nowadays many harbour authorities monitor Channel 16 and their own working frequency. We mention this arrangement later.

Intership Channels

Channel 6 is the primary intership frequency. The regulations state that all ships equipped with VHF radiotelephony must be able to send and receive on Channel 6 — this is in addition to Channel 16 (the VHF Distress Safety and Calling Channel). Channels 8, 9, 10 and 13 have been allocated as intership channels for many years, and now channels 67, 69, 72, 73 and 77 are also available. To communicate with another ship, initial contact may be made on Channel 16 transferring to an intership working frequency for the exchange of traffic. Channels 6, 8, 72 and 77 are exclusively for intership use and should be used in preference to the others. The higher numbered Channels 72 and 77 are used less than Channels 6 and 8. If two ships with these higher numbered channels available wish to communicate they are likely to be less subject to interruption if they use the higher-numbered channels.

Intership channels are not to be used for 'chatting' between ships.

U.K. Small Craft Safety Channel

VHF Channel 67 is available in the U.K. for use by small craft and HM Coastguard for the exchange of *SAFETY* information in situations which do not justify the use of distress or urgency procedures. For example, in bad weather it is permissible to obtain a weather report by calling on Channel 16 and saying you have safety traffic. You will be asked to change to Channel 67 to clear the traffic. As stated in the Notes to Appendix A, Channel 67 may be used for communication between ships, aircraft and land stations for search and rescue co-ordination and for anti-pollution operations.

VHF D/F

Many coastguard stations are now fitted with D/F equipment. The primary purpose of this equipment is to provide bearings of distress calls but the coastguard may be able to provide positional information, using VHF D/F, to any yachtsman who is concerned about his navigational position. It should be stressed, however, that this is not a service to which anyone has an entitlement, it is a bonus if it happens to be available and over-using it would detract from its availability for distress.

"Private" Channels

In addition there are channels available to allow communication with Private Radio Stations. One of these is assigned nationally within the U.K. as the **Marina Channel.** This is Ch.M, 157.85 MHz, and is in use by marinas and yacht clubs in the U.K. which have their own base station. Channel M is virtually a yachtsman's port operations frequency, for communication between a yacht at sea and a marina or yacht club ashore. It may also be used to control club safety boats, or maintain contact between a committee boat and the shore during a regatta. It must not, of course, be used for public correspondence, nor for normal intership communications. The reason for the latter is the danger of causing interference to its authorised use around the U.K. by other yachtsmen or to the services of other countries where the frequency may be used for quite different purposes.

Channel M2(161.425MHz)
This private channel has been allocated as the primary working channel for Yacht Clubs. In cases of overload or non-availability Channel M may be used. Ship Licence holders authorised to operate on Channel M may operate on Channel M2 without further authorisation.

Marina Communication
Channel 80 has been allocated as the primary working channel for Marinas in the United Kingdom. In cases of overload or non-availability Channel M may be used.

HM Coastguard Private Channel
Channel 0, 156,00 MHz is just at the lower limit of the International Maritime Band. Its use is reserved for special purposes, one of which is inter-communication by HM Coastguard. Channel 0 may be fitted in a yacht only if the owner is an Auxiliary Coastguard (Afloat) and authorised by the Department of Trade and Industry to fit and use the frequency. Authorisation to fit channel 0 is not automatically granted to all Auxiliary Coastguards (Afloat).

Channel 70
Channel 70, formerly an intership channel, is now reserved for digital selective calling and must not be used for voice communication.

CHAPTER III

CHOICE OF YACHT EQUIPMENT
AND INSTALLATION

Number of Channels

VHF radiotelephones and VHF monitor receivers for maritime use can be obtained with up to sixty-seven channels (including the private VHF channels outside the International Band).

The modern sets employ a technique known as 'Frequency Synthesis' rather than the traditional plug-in quartz crystal to determine frequencies or channels.

"Dual Watch" facility

Some sets also have the most useful "dual watch" facility. This means that the operator can automatically monitor two channels at the same time without having *manually* to switch from channel to channel. This facility can be usefully employed to monitor, say, a Port Operations frequency and Channel 16 (the International Distress, Safety and Calling frequency).

Older V.H.F. Sets

Certain sets may only have a few channels fitted — particularly the older ones. Hence, if you are contemplating buying a second hand set, or a yacht with a set already installed, it is worth checking that the equipment is good enough for present day usage. If for example, you have to fit a crystal to take the very useful Channel M, you may find (since this frequency is higher than most ship transmitting frequencies) that the set is not capable of accommodating it. With one of the older sets—not with "Frequency Synthesis"—careful planning is required to make sure that its limited capacity is used to the best advantage. A twelve channel set might be set up on the following channels:-

International Distress Safety and Calling	**Channel 16 — mandatory**
Primary Intership	**Channel 6 — mandatory**
Port Operations Service	**Channels 12 and 14**
Small Craft Safety	**Channel 67**

The "Marina" Channel	**Channel M**
Some of the Intership Channels	**8, 72.**
Some of the Ship/Shore Working (Public Corresondence) Channels	**24, 25, 26, 28.**

This set-up will not allow communication with all Coast Radio Stations. If, for instance, the normal sailing area is the Thames Estuary it might be sensible to fit Channel 2 in place of Channel 24, allowing communication with Thames Radio at the expense of Dunkerque. A study of the Admiralty List of Radio Signals, (Vol. 1) or the Nautical Almanacs should indicate which Public Correspondence and Port Operations channels are the most important.

A small three-channel portable set imposes even tighter restrictions.

Simplex or Duplex

Simplex operation is an operating method in which transmission is made possible alternately in each direction. Thus, it is possible either to transmit or receive but not, as on a domestic telephone, to do both simultaneously.

To operate *Simplex* only one aerial is required. To receive, the aerial must be connected to the receiver and to transmit the aerial must be connected to the transmitter. This switching of the aerial from receiver to transmitter and back again is accomplished automatically by means of the 'Press-to-Speak' switch which is normally incorporated in the microphone housing.

Duplex operation is a method in which transmission is possible simultaneously in both directions.

To Work *Duplex,* two frequencies are required. Two aerials or a special duplex filter in the equipment are needed. With this facility normal two-way conversations can be held, in the same way as an ordinary telephone conversation.

Ship-to-shore working Channels are allocated on a two-frequency basis to allow this facility to be used. For example, Channel 26 has two frequencies — the ship transmits on a frequency of 157.3 MHz (shore receives), and the shore station transmits on 161.9 MHz (ship receives).

It is possible to use *Simplex* equipment on the two-frequency channels, remembering that transmission is only possible alternately in each direction. Press to speak: Release to receive.

Semi-Duplex Working is an operating method which is *Simplex* at one end and *Duplex* at the other end of the circuit. As one station has to operate *Simplex,* this method is virtually the same as *Simplex* working, it merely saves the *Duplex* operator having to release his 'Press-to-Speak' switch. Two frequencies are required for this method.

Selcall
This is a facility which allows the yacht to be allocated a number for selective calling by a coast radio station. The call is made by means of a digital code which sounds a signal and switches on a light to indicate that there is a call from a coast radio station.

Autolink
Autolink is a direct dial facility, available as an addition to any existing set. It was introduced in 1991 and allows direct dialling from a yacht at sea, through a UK Coast Radio Station, to a subscriber ashore.

Power Output
The maximum power output permitted for a small-craft VHF radiotelephone transmitter in the Maritime Band is 25 watts. All sets also have facilities for a low power output of about 1 watt. Transmissions should, wherever possible, be made on low power. A low power transmission achieves only short range and thus has a limited possibility of interfering with communication between distant stations.

Position of Radiotelephone
The radiotelephone has to be sited clear of weather so it is usual for it to be in the cabin of a small yacht. In this case, a waterproof extension loudspeaker from the receiver sited close to the steering position is a definite asset, because it allows the watch on deck to monitor the radio without interfering with the sleep of the watch below. Similarly an extension transmitter microphone handset may also be useful.

Power Supplies
The power needed to operate a VHF radiotelephone installation is seldom an important consideration. The specification of a typical 12-volt VHF set suggests that transmitting on 25 watts may consume 5 amperes and that receiving requires only negligible power from the batteries.

Radiotelephone User Controls

A modern VHF radiotelephone is a sophisticated item of equipment but is extremely simple to use. The controls with which the operator must become familiar are no more complicated than the controls of an office dictaphone or a domestic cassette recorder.

Aerials

The choice of a VHF aerial for a particular craft is an important decision. It is a fact that a good transmitting aerial is also a good receiving aerial, but the converse is not necessarily true. Any old aerial may give reception of sorts but the same aerial may prove useless for transmitting.

If the transmitter has an output of 25 watts the **theoretical** aim must be to radiate the full 25 watts from the aerial. Use should be made of the best quality low loss cable, known as 'Aerial Feeder' from the radiotelephone to the aerial.

The height of the aerial is important. The VHF radiotelephone service is known as the 'Short Range Service'. This is because the propagation of radio waves at Very High Frequencies is little more than 'line-of-sight'.

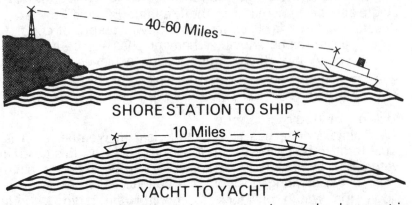

SHORE STATION TO SHIP

YACHT TO YACHT

The radio wave is affected to some degree by barometric pressure, high pressure often giving greater ranges than normal; increased humidity may also increase the range normally attained; and rough seas causing the ship's aerial to sway back and forth over a wide arc often give an effect to the signals known as 'fluttering'.

Two main categories of aerial are available. 'High Gain' aerials are about two metres long and 'Unity Gain' aerials are about one and a half metres long. A "High Gain" aerial has the effect of concentrating the radiated power along an axis

18

at right-angles to the aerial. Thus it makes longer ranges achievable, so long as the aerial is kept nearly vertical as in a motor yacht. In a sailing yacht, constantly "on the heel", the "Unity Gain" aerial is recommended.

It is normally a mistake to site a VHF aerial anywhere other than at a masthead. An aerial sited lower down may be masked by the rigging with the result that communication is difficult with other stations on certain relative bearings.

Installing the Equipment

Fitting a VHF radiotelephone is relatively simple for a handyman and requires no special tools. The set itself should be fastened firmly, in a convenient position which is clear of sea spray and dampness, and away from heat and the engine. An 'earth' is not needed, the set is simply connected up to the yacht's power supply, observing the correct polarity and the aerial feeder cable is plugged in.

The aerial is generally supplied with its own mounting, brackets or plate, and only requires fitting in the highest, most convenient spot.

MF and MF/HF Radiotelephones

The VHF radiotelephone is invaluable for short range contacts but greater range is afforded by a Medium Frequency (MF) or combined Medium Frequency/High Frequency (MF/HF) radiotelephone for those who sail longer distances offshore.

The range which can be achieved using MF from a yacht is likely to be in the order of one hundred miles. Using HF this may be extended to several thousand miles.

The disadvantages of MF/HF radiotelephones are that they are considerably more expensive than VHF (typically 5 to 10 times as expensive) and there are not nearly as many yachts fitted with them. The extra price of MF/HF is probably well worthwhile for anyone sailing across oceans but if a yacht is to cruise or race almost entirely within fifty miles of land the extra expense of MF/HF is difficult to justify.

It should also be noted that a different operator's certificate is required for MF/HF.

Modes of Emission

Before 1970 all MF radio communication was on Double Sideband (DSB). In the early '70s it was internationally agreed that in 1982 DSB communication on all frequencies other than 2182 kHz would be replaced by Single Sideband

(SSB). The purpose of this change was to allow a larger number of frequencies to be used within the same range of the frequency band.

There are still a large number of old DSB sets in existence. They can be used for broadcast reception (BBC and IBA radio stations) and, with a suitable aerial, as direction finding receivers. They may also be used as emergency-only transmitters on 2182 kHz. They cannot, however, receive transmissions from Coast Radio Stations or ships on any other frequency in the MF band and it is generally not an economic proposition to convert them to SSB use.

Emergency Equipment
A VHF radiotelephone is itself an extremely useful item of equipment in an emergency. Distress procedures are described and explained in Chapter VI. There are also specialised items of radio equipment designed specifically for use only in emergencies.

Emergency Radiotelephones
A yacht's main radiotelephone may be disabled in an emergency — as a result of loss of power from the main batteries or loss of the aerial through dismasting. An emergency aerial, which can be rigged on deck after dismasting, is a useful item of equipment to have available. A portable (preferably waterproof) VHF Radiotelephone is also worth considering for use in an emergency. It has the advantages of its own built-in power supply and aerial so it can be used either from a yacht or a life raft. It will not give communication at such a long range as a set with a masthead-mounted aerial and the battery life will be limited. Hence, it should, ideally, be carried in addition to the main radiotelephone.

Emergency only radiotelephones, operating on 2182 kHz are available. These are likely to give slightly longer range than a VHF portable but as they work only on one frequency they are suitable only for emergency use.

Emergency Position Indicating Radio Beacons (EPIRBs)
EPIRBs are available which transmit on one or more of the following frequencies:

121.5 MHz — Civil air distress

243 MHz — Military air distress
406 MHz — Maritime satellite distress

An EPIRB is activated by a vessel in distress and it transmits an emergency signal. Transmissions on 121.5 MHz may be received either by civil aircraft or satellites. Transmission on 243 MHz may be received only by military aircraft and on 406 MHz only by satellites.

An aircraft is likely to have a capability to take bearings or home in on an EPIRB signal. A satellite can obtain a position fix on the transmission, details of which will be transmitted through an earth station to the appropriate rescue co-ordination centre.

The two aircraft distress frequencies have been in use for several decades but the satellite frequency is a recent innovation and allows the EPIRB to transmit a considerable amount of amplifying information, such as the country of origin, the type of the parent vessel and the nature of the emergency.

The problem with EPIRBs has always been the very high rate of false alarms caused by inadvertent activations. The greater sophistication of beacons operating on 406 MHz should go some way to solving this problem.

The most modern beacons operate on all three frequencies. The transmission on 406 MHz raises the alarm and the transmissions on 121.5 MHz and 243 MHz are used for D/F homing by rescue aircraft.

If a beacon is inadvertently activated it must not be switched off until the rescue authorities have been contacted, otherwise a search may continue for many hours to locate the source of the emergency transmission.

Citizen's Band
Citizen's Band radio is not an alternative to Marine Band VHF.

Neither will Citizen's Band allow communication with harbour authorities or Coast Radio Stations.

However, a Citizen's Band radio may, for certain purposes be a useful additional channel of communication between sea-going yachts. It is intended to provide a means of chatting, on any matter (as long as it is not obscene) and one of the commonest **misuses** of Marine Band radio is chatting on intership frequencies. Anyone who wants a radio with which to pass the time of day, or night, gossiping with anyone within range, should fit a CB, preferably in addition to a marine band radiotelephone.

Experience in the USA and Australia, where CB has been in use for many years, has shown that it can also be a useful additional aid to safety. While the Coastguard, very properly, decry CB as no substitute for Marine Band communication, the fact remains that in the next few years CB sets may increase in use. That means that there will be large numbers of people listening to CB frequency, and a call for help may well be picked up by another yacht or an enthusiast ashore. The Coastguard do say that they anticipate CB becoming a useful adjunct to the "999" emergency service.

The essence of CB is that it can be used by anyone and there are no **rules** about procedure. This means that it is widely available and there is no reason why a sailing club should not buy half a dozen sets for use in safety boats. However, the club must then accept that everyone else in the area has a perfect right to use all the CB channels and that no-one can claim the right to use any particular channel for a specific purpose.

CB Licences

Licences for CB Radio are obtainable from Post Offices. The licence conditions cover anybody who lives at the licensee's home, employees of the licensee using sets on his business and anybody who hires sets from the licensee for not more than twenty eight days. One licence fee covers up to three sets and any number of additional fees can be paid to cover extra sets on the same basis.

Citizen's Band is a personal two-way radio service available on two wave bands. 27 MHz FM (VHF) and 934 MHz FM (UHF). 27 MHz is used by the vast majority of CB operators because of the availability of equipment at low cost.

The 27 MHz service can provide forty FM (Frequency Modulated) channels between 27.6 and 28 MHz with a transmitter power of 4 watts. The maximum range is between ten and twelve miles, depending upon terrain.

In the UHF band, (934 MHz FM), there are twenty channels (and likely to be more in the distant future).

CB may not be used to advertise: solicit goods or services: transmit obscene or offensive language: or transmit music and *is essentially a service for speech only.*

In the UK, the 27 MHz AM (Amplitude Modulation) sets are illegal because of the interference which they cause to television, radio, music centres and tape recorders. The introduction of the FM (Frequency Modulated) equipment as the legal version has been effected in the United Kingdom, France, Holland, Germany and the Irish Republic. However seventeen other European countries have CB where AM is legal.

AM sets in the United Kingdom can be modified to meet the specification of the new FM service, but the modification is not always easy.

The sets which conform to the "legal" service in the UK must all carry on the front panel a circle containing either the legend CB 27/81 or CB 934/81.

Procedure on CB radio

The Home Office, in consultation with other bodies, has published a short code of practice which is worth the reader's attention. It does not vary much from the ordinary required procedure for operating VHF equipment in the Maritime Mobile Band. **On Citizens Band:** —

Leave Channel 9 clear for emergencies. Whilst there is no official monitoring of this channel it has already proved a useful method of obtaining voluntary assistance from others. Incidentally, if there is no answer on Channel 9 you may call for help on either Channel 14 or 19.

Channel 14 is the general calling channel and once you have established a contact you should move to another channel to hold your conversation.

Channel 19 is for "main road" users and is the one most frequently used by long distance transport drivers.

To avoid interference problems, position your antenna as far away as possible from others and remember that you are **not allowed to use power amplifiers.**

Always listen before you transmit (with the "squelch" control turned fully down and "tone squelch" turned off if you have selective call facilities) to ensure that you will not be transmitting on top of any existing conversation.

Keep conversations and transmissions short and listen often for a reply since you may find that the station with which you are corresponding has moved out of range.

Always leave a short pause before replying.
Plain language is just as effective as CB slang.
Be patient with newcomers and assist them.

CHAPTER IV

WHAT, HOW AND WHEN TO SAY IT

Procedure Cards

Ships and fishing vessels required by law to fit radiotelephones are required to display cards, which can be read easily from the radiotelephone operation position, setting out the Distress, Urgency and Safety message procedures. Yachts which fit radiotelephones voluntarily are not bound to display these cards but in a distress situation the skipper and other members of the crew may be put out of action and the lives of survivors may depend on their ability to send a distress message correctly. Everyone on board a yacht should know how to operate the radiotelephone.

Similarly any member of the crew may receive a Distress, Urgency or Safety call and other peoples' lives may depend upon the correct action being taken.

Appendix E gives an example of what *might* be displayed to assist the totally inexperienced operator to send a Mayday Signal.

On most VHF radiotelephones the operator MUST press a switch in the hand set to speak and RELEASE IT to listen. To most of our readers this is probably already known. **We emphasise it because lives have been lost because an inexpert operator did not know this.** It is sense therefore to put the simplest instructions on a card near the set explaining at least what to do to call for help. Since nearly every set has different controls it is almost a question of labelling everything and setting out a step-by-step 'idiot's guide to making the thing work'.

It is usually too late when an inexperienced operator has to bend down to read the little slogans above each knob. Those who own sets (and yachts) are well advised to work out the simplest method of organising 'Procedure Cards' for their own vessel.

If the radio is fitted with dual watch, there is a danger of assuming that it is switched to Channel 16, because channel 16 traffic has been received. There have been instances of distress messages being transmitted on un-guarded working frequencies because the operator, having been listening to typical Ch 16 transmissions, on dual watch, wrongly assumed that the set was switched to Ch 16 and did not make a positive check before sending the distress call.

The Purpose of Standard Procedure
English is one of the recognised international languages of radiotelephony, but accents can easily make words difficult to distinguish, and radio interference can make the clearest voice difficult to understand. Standard procedure and familiar words and phrases provide a common pattern, understood by radio-operators of most nationalities.

When the standard words or phrases are used in an expected order they are much easier to discern against a background of radio and weather interference. Departures from the standard procedure often create confusion, reducing the reliability and speed of communication. The correct procedure is well worth learning.

Transmission Rules on Maritime Frequencies
The following simple rules are essential to the efficient use of the radiotelephone frequencies and channels. In most cases they are necessary to conform with international rules of conduct.

The following are strictly forbidden:
1.   Transmissions which have not been authorised by the Master or other person in charge of the ship.
2.   Operation of a radiotelephone by unauthorised persons. Passengers or other members of the crew may make radiotelephone calls under supervision.
3.   The transmission or circulation of false or deceptive distress, safety or identification signals.
4.   Transmissions made without identification, i.e. without ship's name or call-sign.
5.   The use of christian names or other unauthorised identification in lieu of ship's name or call-sign.
6.   Closing down a radiotelephone before finishing all operations resulting from a distress call, urgency or safety signal; exchanging all traffic on hand (or indicated) with the Coast Radio Station or other ships which have indicated a desire to communicate with you.
7.   Broadcasting messages or programmes. To "broadcast" means to transmit (without a reply being expected) information intended for reception by another person or

persons. (Broadcasting safety messages to "All Ships" is an exception to this rule).

8. **Making unnecessary transmissions or transmitting superfluous signals.**

9. The transmission of profane, indecent or obscene language.

10. The use of frequencies or channels other than those covered by the ship's licence.

11. The broadcast transmission of music.

12. The broadcast of messages intended for reception of addresses on shore except through a Coast Radio Station.

Secrecy of Correspondence
Radio operators and others who become acquainted with the contents of radiotelegrams or radiotelephone calls are bound to preserve the secrecy of correspondence. No one shall divulge the contents or even the existence of correspondence transmitted, received or intercepted by a radio station.

The interception of radio communication correspondence, other than that which the station is authorised to receive, is forbidden and in the case where such correspondence is involuntarily received, it must not be reproduced, nor communicated to third parties, nor used for any purpose, and even its existence shall not be disclosed.

Avoidance of Interference
Before transmitting, first listen on the frequency or channel to make sure that your transmission will not interfere with any other communications already in progress. If the frequency is occupied, then wait for a break before transmitting.

Having taken this precaution, if you do cause interference you must comply with any request from a Coast Radio Station to stop transmitting. The request will contain an indication of the time for which you should refrain from transmitting.

Control of Communications
Ship-to-Shore: Except in the case of distress, urgency or safety, communications between a ship and a Coast Radio Station are controlled by the Coast Radio Station.

Intership: The ship which is called controls communication. If you call another ship, then that ship has control, or, if you are called by a ship, you must assume control. If a Coast Radio Station finds it necessary to interrupt, both ships must comply with any instructions given by the Coast Radio Station.

It must be borne in mind that a Coast Radio Station generally has much better aerials and equipment than most ship stations and thus the area covered by its transmissions and reception is greater.

The diagram explains the reasons for these rules. Both ship stations are able to communicate with the Coast Radio Station but not with each other. Ship station 'B' could be transmitting; Ship station 'A', not able to hear 'B', thinks the Channel is clear and transmits. The Coast Radio Station receives both signals as interference, and requests 'A' to stop transmitting while he deals with 'B'. When the channel is clear, the Coast Radio Station will call Ship station 'A'.

COAST RADIO STATION

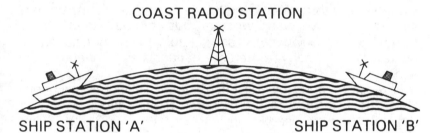

SHIP STATION 'A'   SHIP STATION 'B'

Voice Technique

Operating a radiotelephone is an art in which personality plays a full part. There are two basic considerations when voice operating:

WHAT to say (e.g. voice procedure)

HOW to say it (e.g. voice technique)

The necessity for clear speech on a radiotelephone cannot be over-stressed. If a message cannot be understood by the receiving operator it is useless.

Almost anyone can become a good radiotelephone operator by following a few simple rules:-

Pitch. The voice should be pitched at a slightly higher level than for normal conversation. Any tendency to drop the pitch of the voice at the end of a word or phrase should be avoided.

Volume. The microphone should be held a few inches in front of the mouth and spoken directly into, at normal conversation level.

Clarity. Speak clearly, so that there can be no confusion with other words; words with weak syllables should be emphasised (e.g. 'Tower', if badly pronounced, could sound like 'tar'.) People who have a particularly strong accent must try to use as understandable a pronunciation as possible.

Speech rate. Messages which have to be written down (copied) by the receiving station should be sent slowly. They should be spoken in natural phrases, not word by word and a pause made at the end of each phrase to allow sufficient time for it to be written down by the receiving operator. Remember, average *reading* speed is two hundred and fifty words a minute, whilst average writing speed is only twenty words a minute.

The Standard Phonetic Alphabet

Whilst there are, and have been, numerous ways of identifying with certainty the letters of the alphabet, the one set out below was worked out by an International Committee as being the most suitable for pronunciation by operators of many different nationalities, tongues and accents.

It is recommended by the International Telecommunications Union (ITU) for use on the Maritime Mobile Bands whenever radiotelephony is used.

The syllables to be emphasised are printed in bold type. Incidentally if you use your radiotelephone infrequently it is a good idea to have this alphabet exhibited near the set for your ready use. To lapse into such fatuities as "L for leather" only displays your comparative amateurish-ness!!

Letter	Word	Spoken as	Letter	Word	Spoken as
A	ALFA	**AL**FAH	N	NOVEMBER	NO**VEM**BER
B	BRAVO	**BRAH**VOH	O	OSCAR	**OSS**CAR
C	CHARLIE	**CHAR**LEE	P	PAPA	PAH**PAH**
D	DELTA	**DELL**TAH	Q	QUEBEC	KEH**BECK**
E	ECHO	**ECK**OH	R	ROMEO	**ROW**MEOH
F	FOXTROT	**FOKS**TROT	S	SIERRA	SEE**AIR**RAH
G	GOLF	**GOLF**	T	TANGO	**TANG**GO
H	HOTEL	HOH**TELL**	U	UNIFORM	**YOU**NEEFORM
I	INDIA	**IN**DEEAH	V	VICTOR	**VIK**TAH
J	JULIETT	**JEW**LEEETT	W	WHISKEY	**WISS**KEY
K	KILO	**KEY**LOH	X	X-RAY	**ECKS**RAY
L	LIMA	**LEE**MAH	Y	YANKEE	**YANG**KEY
M	MIKE	**MIKE**	Z	ZULU	**ZOO**LOO

Difficult words, or groups of letters within the text of messages may be spelled using this Phonetic Alphabet. The operator will precede this spelling with the words 'I SPELL'. If the word to be spelt happens also to be pronounceable, it should be included both before and after it has been spelt.
Example:-
'INTEND ANCHORING OFF YOUGHAL – I SPELL – YANKEE OSCAR UNIFORM GOLF HOTEL ALFA LIMA – YOUGHAL–'
Another example might be where a Coast Radio Station asks for your International Callsign. If your Callsign were MBDD, you would transmit:-
'MY CALLSIGN IS MIKE BRAVO DELTA DELTA.'

Phonetic Numerals
When numerals are transmitted by radiotelephone, the following rules for their pronunciation should be observed.

Numeral	Spoken as
1	WUN
2	TOO
3	TREE
4	**FOW**-ER
5	FIFE
6	SIX
7	**SEV**-EN
8	AIT
9	**NIN**-ER
0	ZERO

Numerals should be transmitted digit by digit except that multiples of thousands may be spoken as such.

Numeral	Spoken as
44	**FOW**-ER **FOW**-ER
90	**NIN**-ER ZERO
1478	WUN **FOW**-ER **SEV**-EN AIT
7000	**SEV**-EN THOUSAND
136	WUN TREE SIX
500	FIFE ZERO ZERO

Procedure Words (Pro-words)
Pro-words might be described as 'professional' words! What the professionals say. All these words are designed for easy international understanding and **BREVITY. NOT** included are such rambling pieces of chat as 'I am receiving you loud and clear' or antiquated pieces of World War usage such as

'Wilco'. The Standard Marine Navigational Vocabulary contains a number of the correct words to use.

The list below is compiled from that publication, and from RNLI and H.M. Coastguard Instruction books.

ALL AFTER — Used after the proword 'SAY AGAIN' to request a repetition of a portion of a message.

ALL BEFORE — Used after the proword 'SAY AGAIN' to request a repetition of a portion of a message.

CORRECT — Reply to a repetition of a message that has been preceded by the prowords 'READ BACK FOR CHECK', when it has been correctly repeated.

CORRECTION — Spoken during the transmission of a message means — An error has been made in this transmission. Cancel the last word or group. The correct word or group follows.

IN FIGURES — The following numeral or group of numerals are to be written as figures.

IN LETTERS — The following numeral or group of numerals are to be written in letters as spoken.

I READ BACK — If the receiving station is doubtful about the accuracy of the whole or any part of a message it may repeat it back to the sending station, preceding the repetition with the prowords 'I READ BACK'.

I SAY AGAIN — I am repeating transmission or portion indicated.

I SPELL — I shall spell the next word or group of letters phonetically.

OUT — This is the end of working to you. The end of work between two stations is indicated by each station adding the word OUT at the end of its last reply.

OVER — The invitation to reply. Note that the phrase "OVER AND OUT" is **never** used.

RADIO CHECK — Please tell me the strength and the clarity of my transmission.

RECEIVED — Used to acknowledge receipt of a message, i.e. 'YOUR NUMBER . . . RECEIVED'. In cases of language difficulties, the word 'ROMEO' is used.

SAY AGAIN — Repeat your message or portion referred to i.e. 'SAY AGAIN ALL AFTER . . . SAY AGAIN ADDRESS' etc.

(Note:- This is typical of the need to memorise all these 'catch phrases'. To use the word 'Repeat' would be wrong. 'Repeat' is used to emphasise something.)

STATION CALLING — Used when a station receives a call which is intended for it, but is uncertain of the identification of the calling station.

THIS IS — This transmission is from the station whose callsign immediately follows. In cases of language difficulties the abbreviation **DE** spoken as 'DELTA ECHO' is used.

WAIT — If a called station is unable to accept traffic immediately, it will reply to you with the prowords 'WAIT . . . MINUTES'. If the probable duration of the waiting time exceeds 10 minutes the reason for the delay should be given.

WORD AFTER or **WORD BEFORE** — Used after the proword 'SAY AGAIN' to request a repetition of a portion of a radiotelegram or message.

WRONG — Reply to a repetition of a radiotelegram that has been preceded by the prowords 'I READ BACK', when it has been incorrectly repeated.

The Standard Marine Navigational Vocabulary (M1252)

The Standard Marine Navigational Vocabulary was compiled to standardise the language used in communication for navigation at sea, in port approaches, waterways and harbours.

The vocabulary sets out certain standard phrases and terms which should be recognised internationally. Extracts of the more important sections of Merchant Shipping Notice M1252 (available from Mercantile Marine Offices, Customs Offices and Harbour Offices) are set out below.

The notice also contains an excellent Glossary of terms and you are highly recommended to obtain a copy.

Repetition

If any parts of the message are considered sufficiently important to need safeguarding, use the word "repeat", eg
"You will load 163 — repeat — 163 tonne bunkers,"
"Do not — repeat — do not overtake".
In this case, and in the examples of messages given elsewhere, all the numerals should, of course, be spoken phonetically — as previously described.

Position

When latitude and longitude are used, these shall be expressed in degrees and minutes (and decimals of a minute if necessary), North or South of the Equator and East or West of Greenwich.

When the position is related to a mark, the mark shall be a well-defined charted object. The bearing shall be in the 360° notation from True North and shall be that of the position *FROM* the mark.

Examples: "THERE ARE SALVAGE OPERATIONS IN POSITION ONE FIVE DEGREES THREE FOUR MINUTES NORTH SIX ONE DEGREES TWO NINE MINUTES WEST".
"YOUR POSITION IS ONE THREE SEVEN DEGREES TWO POINT FOUR MILES FROM BARR HEAD LIGHTHOUSE".

Courses
Always to be expressed in 360° notation from North (true North unless otherwise stated). Whether this is *to* or *from* a mark can be stated.

Bearings
The bearing of the mark or vessel concerned, is the bearing in the 360° notation from North (true North unless otherwise stated), except in the case of relative bearings.
However, bearings may be either FROM the mark or FROM the vessel.

Examples: "THE PILOT BOAT IS BEARING TWO ONE FIVE DEGREES FROM YOU".
"YOUR BEARING IS ONE TWO SEVEN DEGREES FROM THE SIGNAL STATION".

Note: Vessels reporting their position should always quote their bearing *FROM* the mark.

Relative Bearings
Relative bearings can be expressed in degrees relative to the ship's head/bow. More frequently this is in relation to the port or starboard bow,

Example: "THE BUOY IS THREE ZERO DEGREES ON YOUR PORT BOW". (However, relative DF bearings are more commonly expressed in the 360° notation).

Distances
Preferably to be expressed in nautical miles or cables (tenths of a mile) otherwise in kilometres or metres, the unit always to be stated.

Speed
To be expressed in knots
(a) without further notation meaning speed through the water; or
(b) "ground speed" meaning speed over the ground.

Numbers
Numbers are to be spoken thus "One-Five-Zero" for 150. "Two point five" for 2.5.

Geographical Names
Place names used should be those on the chart or Sailing Directions in use. Should these not be understood latitude and longitude should be given.

Time
Times should be expressed in the 24 hour notation indicating whether GMT, zone time or local shoretime is being used.

Test Calls
Test calls may be made to check that the equipment is functioning correctly. The duration of a test call must not exceed 10 seconds.

CHAPTER V

CALLS AND CALLING AND
PUBLIC CORRESPONDENCE

Callsigns

Coast Radio Stations normally identify themselves by using their geographical name followed by the word RADIO, e.g. Niton Radio, Humber Radio, Lands End Radio, etc.

Ship stations normally identify themselves by the name of the ship as shown on their Ship Station Licence but the International callsign assigned to the ship when the licence is issued may be used in certain cases. If there are two or more yachts bearing the same name or where some confusion may otherwise result, you should give your International callsign when establishing communications, and thereafter use your ship's name as the callsign.

TRANSMISSIONS WITHOUT IDENTIFICATION ARE FORBIDDEN.

'All Ships' Broadcast

Information intended to be received or used by anyone who can intercept it, e.g. Gale Warnings, Navigational Warnings, Weather Forecasts, etc., is generally broadcast by Coast Radio Stations and addressed to 'ALL STATIONS'. No reply is to be made to this type of broadcast.

Priority of Radiotelephone Calls

Radiotelephone calls are subject to a priority list, as shown below:

1. Distress.
2. Urgency.
3. Safety.
4. Radio direction-finding.
5. Navigational and safety movement of aircraft.
6. Navigation, movements, and needs of ships, and weather observation messages destined for an official meteorological service.
7. Government radiotelegrams relative to the application of the United Nations Charter. These telegrams bear the prefix ÉTATPRIORITÉNATIONS.
8. Government radiotelegrams with priority bearing the prefix ÉTATPRIORITÉ or ÉTAT and Government calls for which priority has been expressly requested.

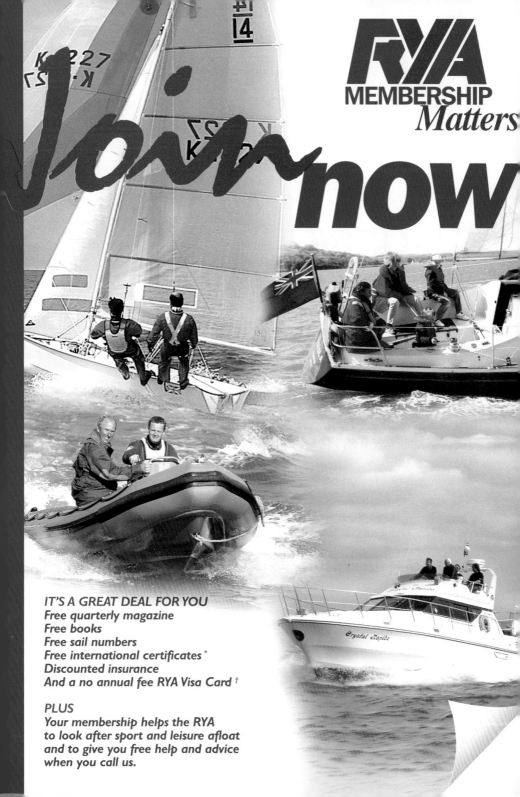

RYA
MEMBERSHIP *Matters*

Join now

IT'S A GREAT DEAL FOR YOU
Free quarterly magazine
Free books
Free sail numbers
Free international certificates *
Discounted insurance
And a no annual fee RYA Visa Card †

PLUS
Your membership helps the RYA
to look after sport and leisure afloat
and to give you free help and advice
when you call us.

RYA
MEMBERSHIP
Matters

It's a great
deal for you...

Membership costs very little...
- Personal Membership £25 (or £23 by direct debit)
- Under 21 £10
- Family Membership £40 (or £38 by direct debit)

and there are FREE BENEFITS...
- Four editions of RYA News every year
- RYA publications
- Allocation of sail numbers
- A no annual fee RYA Visa Card[†]
- International certificates of competence [*]
- Boat show lounges

special discounts on
- Boat insurance

...and more
As an RYA member you can talk directly to us and get information and legal advice on all aspects of leisure boating from inland waters cruising to Olympic racing.

You can benefit from all this expertise by joining now... then simply pick up the phone the next time you need advice.

[†]Only available to UK residents aged 18 years and over and subject to status.   [*]If you qualify

and important for sport and leisure afloat

So, those are the benefits, but what happens to your subscription?

Your subscriptions enable us to work for the interests, and protect the freedoms of, *everyone* who enjoys boating.

Imagine a world where the RYA did not work with UK and EU legislators to protect your freedom to cruise the inland waterways, estuaries and seaways as you always have. Suppose there was nobody setting minimum safety standards and providing high quality training schemes for schools and instructors. Imagine no one working to attract people into the sport, or developing the potential of talented young sailors and powerboat racers. Imagine not having a strong team to represent us at the Olympic Games, or a collective voice to improve cruising facilities and marina standards.

It's the RYA that does all these things, and more – and it's your membership that makes it happen.

please join now

Your membership matters. It's important for everyone who enjoys their sport and leisure afloat - and it's a great deal for you.
There is a membership form overleaf for you to use

Yes I want to join the RYA

Type of Membership Required: (tick as applicable)

☐ **Personal £25** (*£23* if you pay by Direct Debit)

☐ **Family £40** (*£38* if you pay by Direct Debit)

☐ **Under 21 £10**

Please indicate your main boating interest by ticking one box only	W	SC	SR	PR	MC	PW	
	☐	☐	☐	☐	☐	☐	W=Windsurfing SC=Sail Cruising SR=Sail Racing PR=Powerboat Racing MC=Motor Cruising PW=Personal Watercraft

	Title	Forename	Surname	Date of Birth	Male	Female
1.						
2.						
3.						
4.						

Address

Town   County   Postcode

Signature _

THE EASY WAY TO PAY

INSTRUCTIONS TO YOUR BANK OR BUILDING SOCIETY TO PAY DIRECT DEBITS

Originators identification number

9	5	5	2	1	3

DIRECT Debit

Please complete this form and return it to:
Royal Yachting Association RYA House Romsey Road Eastleigh Hampshire SO50 9YA. Tel: 01703 627400

To The Manager _____

_____ **Bank or Building Society**

Address _____

_____

_____ **Post Code** _____

2. Name(s) of account holder(s)

3. Branch Sort Code

| | | | — | | | — | | |

4. Bank or Building Society account number

5. RYA Membership Number (For office use only)

6. Instruction to pay your Bank or Building Society

Please pay Royal Yachting Association Direct Debits from the account detailed on this instruction subject to the safeguards assured by The Direct Debit Guarantee

Signature(s) _____

Date _____

Cash, Cheque, Postal Order
enclosed £ _____ Made payable to the Royal Yachting Association

Office use / Centre Stamp

077 **Office use only:** Membership No. Allocated

9. Service communications relating to the working of the telecommunication service or to communications previously exchanged.

10. All other communications, i.e. radiotelephone calls.

Calling a Coast Radio Station

The working frequencies of coast radio stations are listed in a number of different publications. The Yachtsmen's Nautical Almanacs provide a useful list.

Select a working frequency and listen for several seconds to ensure that it is free for use. If it is in use you will hear either someone talking or a series of pips.

The initial call to a CRS is the only one in which great brevity cannot be effective because a call of at least three seconds is required to switch on equipment at the CRS and register your call. Thus it is sensible to repeat the name of the CRS twice and your own callsign twice in the initial call. Coast Radio Stations tend to be very busy at times and it is quite common for a CRS to have four VHF channels, a couple of MF radiotelephone channels and an MF wireless telegraphy (morse) channel, all manned by just two operators. There may occasionally be a short delay before your initial call is answered by an operator but if that call is on a working channel you should immediately hear the channel engaged signal — this means that your call has activated the automatic equipment in the CRS and an operator will answer you as soon as he is free to do so.

So, wishing to raise Niton you listen carefully before transmitting.

If there is silence then you can call:—

NITON RADIO, NITON RADIO – THIS IS BARBICAN, BARBICAN – CALLSIGN MIKE MIKE ALPHA ALPHA, MIKE MIKE ALPHA ALPHA – ONE LINK CALL PLEASE – OVER.

you will then hear the "answering" pips indicating that your call has been registered and the operator will answer you as soon as possible.

Don't forget that if you don't hear anything your set may not be working. Whatever the cause wait three minutes before starting again.

Charges

After the call is finished the Coast Radio Station will decide the duration of the call for charging purposes and will normally inform you of the duration immediately after the end of the call.

The end of the call, for timing purposes, is registered when the subscriber ashore puts his telephone down. If the operator in the CRS does not immediately tell you the time for which you will be charged don't worry, you will not be charged for extra time after the call has finished.

There are three ways of paying for radiotelephone link calls. You may be sent an account as the owner of the yacht making the call, you may charge the call to your home telephone number (a system known as YTD) or you may make a reverse charge call. There is an extra payment equivalent to two minutes for reverse charges.

Delays
If a Coast Radio Station has any difficulty in establishing a shore connection the yacht will be informed. She should continue to listen until a connection can be established or she may arrange to call the Coast Radio Station later on.

Calls to yachts from shore
Incidentally, Link Calls can, of course, be made from shore to ship. In that event the operator will call you on Channel 16 and if he receives no reply he will add your vessel's name to his traffic list which is broadcast at certain times each day (see Traffic Lists).

Contacting another ship
Communication with another yacht or ship is simple. Unless you have already made a special arrangement the initial call is on Channel 16. As soon as communication is established it is up to the vessel being called to nominate the intership working frequency to be used. The initial call should be as brief as possible so that Channel 16 is not cluttered up:—
HONEYBEE – THIS IS BARBICAN, BARBICAN – OVER.

That is probably the shortest call which will work but, if the other ship is expecting you to call, you need only use your own yacht's name once.

After, on the nomination of the ship called, you have moved to a working frequency there is less need for brevity but that does not mean that it is permissible to monopolise an intership channel for half the afternoon. The number of channels available is limited and a number of people may want to use them. They are not for general chatter, they are for passing messages concerned with the business of vessels.

It is often assumed that intership VHF radiotelephony is likely to be a useful aid to collision avoidance because it allows two approaching vessels to speak to each other. The problem is that the two approaching vessels may be uncertain of each others' identities and there is a risk of apparently agreeing to manoeuvre in a particular way, when in fact one side of the agreement was given by a third vessel and one of the approaching vessels remains totally unaware that any exchange of information by R/T has taken place. Even when it is possible to be sure of the identity of each vessel, experience has shown that the time taken in establishing radio contact and confirming identities is so long that the use of radio is sometimes a factor contributing to, rather than avoiding, a collision.

Contacting a yacht from the shore

In the first place it is sensible to give your friends and relatives the telephone numbers of Coast Radio Stations which may be able to contact you whilst you are at sea. The telephone numbers are given in Yachtsmens' Nautical Almanacs.

Also shown in the Post Office Telephone Dialling Code books is a heading under 'Maritime Services'. Should the telephone number of the appropriate Coast Radio Station not be known, your correspondent can dial '100' and ask for "Ships Telephone Service".

Traffic Lists

Traffic arriving at Coast Radio Stations for ships at sea within their range is transmitted to those ships in the following way. The Coast Radio Station will make up a "Traffic List" consisting of the names or callsigns of all ships for which traffic is held. These lists are transmitted on normal working frequencies of the Coast Radio Station at definite times. The times and working frequency for each Coast Radio Station are shown in reference books such as Nautical Almanacs.

The broadcast of a Traffic List is preceded by an announcement on the calling frequency, Channel 16, in the following form:

On Channel 16:

HELLO ALL SHIPS – THIS IS SEVERN RADIO – SEVERN RADIO – LISTEN FOR MY TRAFFIC LIST ON CHANNEL TWO FIVE – OUT.

Ships then shift to the working channel and listen for the traffic list. Severn Radio transmits on Channel 25:

HELLO ALL SHIPS – THIS IS SEVERN RADIO – SEVERN RADIO – I HAVE TRAFFIC ON HAND FOR THE FOLLOWING SHIPS – ALFRESCO – CAMPION – GOLF ALFA BRAVO ECHO –

(and so on)

Ships should endeavour to listen to the traffic lists transmitted by Coast Radio Stations, and upon hearing their name or other identification in the list, contact the Coast Radio Station as soon as the list is completed.

If a Coast Radio Station requires to contact a ship between traffic list times it will call the ship on Channel 16.

Calling Harbour Authorities

Most harbour authorities nowadays listen out on "dual watch". They will probably be monitoring two frequencies, Channel 16 and their own allocated frequency on a Port Operations Service frequency. Only traffic relating to Port Operations can be conducted on such a frequency.

These harbour authority frequencies are listed in yachtsmen's Nautical Almanacs and in Volume 6 of the Admiralty List of Radio Signals.

It is nowadays permissible to avoid using Channel 16 and to call the harbour authority direct on its working frequency but, NOT on Channel M because that is not a port operations frequency.

Obviously, if a harbour authority operator is listening out on dual watch, he would not know which channel was being used for your call unless he actually looked at his equipment at the moment of the call. Thus, to avoid any ambiguity, the practice has developed of calling thus: —

HARWICH HARBOUR RADIO – THIS IS JASMINE, JASMINE – ON CHANNEL TWELVE – OVER

If you pass your message in this way the Harwich operator knows (without looking) which channel is being used to call him.

Again note the brevity of this message.

Garbled Calls
Station called — callsign garbled

When a station receives a call without being certain that it is intended for it, it *must not reply* until the call has been repeated and understood.

e.g. Imagine that you heard a transmission, the first part of which was unclear but you could hear the second part of the call:—

THIS IS HUMBER RADIO – HUMBER RADIO – HAVE A LINK CALL FOR YOU – OVER

You must bear in mind that other ships listening on the frequency or channel would also have heard this garbled call, and if they all replied the result would be chaos! Clearly, if Humber Radio (for example) gets no reply to a call, he will repeat it, until he establishes contact with the correct station.

Calling station-callsign garbled

When a station receives a call which is intended for it, but is uncertain of the callsign of the calling station, it should reply as follows:—

STATION CALLING BARBICAN – STATION CALLING BARBICAN – THIS IS BARBICAN – SAY AGAIN – OVER

Unanswered Calls

Before repeating a call, check that the controls on your set are correctly adjusted — power on, high power selected, volume turned up, squelch turned down and correct channel selected. Continued repeated calls are a frequent source of unnecessary use of a channel. They very often result from the calling station being unable to hear the answer to the initial call, either because the set is not correctly adjusted or the "press-to-transmit" switch is jammed in the transmit position. You **must** wait a minimum of three minutes before repeating a call, except when sending a Distress Call.

CHAPTER VI

DISTRESS

International Nature of Distress Procedures
One of the great advantages afforded by a radiotelephone on a yacht or other small craft is the ability to call for assistance if you are involved in a Distress or Urgency situation.

The Maritime Mobile Radiotelephony Bands are International, and at no time does this become more evident than when a ship in distress calls for assistance. By taking part in this international organisation and following the correct procedures, you can help to ensure that you or any other ship in distress obtains assistance without delay.

Distress Frequencies
Although this booklet is primarily concerned with VHF Radiotelephony, the frequency 2182 kHz is the International Distress Safety and Calling frequency for radiotelephony in the MF band. All U.K. and many foreign Coast Radio Stations keep continuous watch on this frequency and all ships fitted solely with MF should keep the maximum practicable watch on this frequency whilst at sea. When a continuous watch is not practicable, ships should make every effort to listen on 2182 kHz twice each hour for 3 minutes commencing at the hour and at the half hour. During these 3 minute "Silence Periods" all transmissions on the frequency except for Distress and Urgency communications must cease.

On VHF, Channel 16 (156.8 MHz) as already mentioned is the International Distress, Safety and Calling frequency.

All Coastguard stations and all U.K. and most foreign Coast Radio Stations keep continuous watch on VHF Channel 16. Ships fitted solely with VHF radiotelephones should maintain the maximum practicable watch on Channel 16 when at sea. There are no Silence Periods on Channel 16.

Exchange of calls on Channel 16 should not exceed one minute unless they are for Distress, Urgency or Safety reasons. Remember again the need for brevity and for switching to another working channel for normal traffic.

Remember also that, in less than distress or urgency situations, the U.K. Small Craft Safety Channel (Channel 67) can be used to speak to HM Coastguard.

It should not go without comment that in case of difficulty any frequency may be used for genuine emergency if the usual distress channels are unavailable for some reason. The primary inter-ship channels 6 and 8 are likely to be the most used channels upon which someone will hear you.

The words **Distress, Urgency** and **Safety** have, in the context of radiotelephony, very special meanings.

DISTRESS

DISTRESS is announced by the word "Mayday" (which derives from the French language *M'aidez* — Assist me) and indicates that a **ship or person is threatened by grave and imminent danger and requests immediate assistance.** Its use for less urgent purposes may result in insufficient attention being paid to calls made from ships which require such immediate assistance. Less urgent signals are dealt with in Chapter VII.

Technically, there are two distinct elements in the transmission of a distress signal and they are separately described below. However, the distress call should always be followed, immediately, by the distress message, so in practice the two elements are not separated.

The Distress Call

This call has absolute priority over all other transmissions. All stations hearing it must immediately cease any transmissions which could cause interference to the distress traffic. They must continue to listen on the frequency for the distress message which follows. The RYA has been advised that despite the need for speed it is helpful to indicate that the call comes from a yacht.

— distress signal	MAYDAY – MAYDAY –
(three times)	MAYDAY –
— the prowords	THIS IS –
— the callsign of the ship	YACHT CATSPAW – YACHT
(three times)	CATSPAW – YACHT CATSPAW.

Distress Message

The distress message which comes after a distress call should follow a laid down procedure but should pack in as much information as possible.

The position of the vessel in distress should be given in

41

terms of latitude and longitude, or whenever practicable, as a TRUE bearing and distance FROM a known geographical point. If, however, the vessel in distress is on a rock or shoal, or near a headland or other identifiable place, a precise geographical indication of the position should be given. When drifting, state the direction and rate of drift. An indication of any subsequent change in the position or intentions of those on board should also be given.

The message: —

distress signal	MAYDAY
name of the ship	YACHT CATSPAW
position	ONE EIGHT ZERO
	FROM CALDY ISLAND ONE MILE
nature of distress	STRUCK ROCK – DISABLED
	AND IN SINKING CONDITION
assistance required	REQUIRE IMMEDIATE
	ASSISTANCE
other information to help	TOTAL CREW FOUR ON
rescue ships	BOARD BLUE MOTOR YACHT
	DRIFTING SOUTH WEST
invitation to reply and	OVER
acknowledge	

The sender may expect an *immediate* reply or acknowledgement on the channel used from another ship, a Coast Radio Station or the Coastguard. If this is *not* forthcoming, check the equipment and repeat the Distress Call and Distress Message, at regular intervals.

If no answer is received, the message may be repeated on any other available frequency or channel upon which attention might be attracted.

Obligation to Acknowledge Receipt of a Distress Message

The International Rules state, 'The obligation to accept distress calls and messages is absolute in the case of every station without distinction, and such messages must be accepted with priority over all other messages; they must be answered and the necessary steps must immediately be taken to give effect to them'.

Let us see what the rule means to a yacht fitted with a radiotelephone. There are a number of factors to consider:

1. You hear the distress message on Channel 16 VHF. Propagation conditions at VHF are such that the vessel in distress could be anywhere within a fifty mile radius of your receiver.

2. The position given in the distress message (if correct) would tell you whether you were in vicinity of the ship in distress or not.

3. The name, callsign or other identification used by the ship in distress may indicate the size of ship and the possible number of persons on board.

4. With the information available to you, are you in a position to render assistance to the station in distress?

If you acknowledge receipt of the distress message, you are in fact telling the ship in distress that you have received her message and may be proceeding to her assistance.

5. A station in distress more than fifty miles from the coast will probably obtain speedier assistance from other shipping in the vicinity than from shore-based rescue facilities. It is desirable that the distress message reaches a Coast Radio Station for there it will be retransmitted to All Ships by the generally more powerful Coast Radio Station transmitters on all distress channels. Also, the Coast Radio Station has immediate access to the Coastguard who will alert the Search and Rescue organisations, i.e. aircraft and helicopters and ships at sea.

6. On the other hand, a station in distress within fifty miles of the coast, although requiring any shipping in her vicinity to provide assistance, also wants the assistance of a Coast Radio Station and Coastguard, both to retransmit her distress message and to alert the Search and Rescue organisations such as the RNLI, SAR aircraft and helicopters.

7. **The decision to acknowledge a distress message is a very important one.** On receipt of a distress message from a mobile station, if you are in an area where reliable communication with a Coast Radio Station is practicable, wait a few moments to allow the Coast Radio Station or the Coastguard to acknowledge receipt, or to allow ships nearer to the vessel in distress to acknowledge receipt.

8. If no acknowledgement is forthcoming, there are two possible courses of action open to you. Clearly a small yacht cannot render much assistance to an ocean-going passenger liner but if the distress message were from a small yacht and you were in a position to render assistance **then** you could set about going to help and would acknowledge receipt. If

not in a position to render assistance, you must take all possible steps to contact a Coast Radio Station, the Coastguard or other vessels which might be able to help. The Distress Call and the Distress Message must be re-transmitted under the rules for **MAYDAY RELAY.**

Not in the Vicinity
Listen for stations nearer to the mobile station in distress to acknowledge receipt. If no acknowledgement is forthcoming, re-broadcast as a **MAYDAY RELAY.**

Always write down a distress message
Always write down a distress message, or if it is being transmitted too quickly to write down all of it at least note the name of the vessel in distress and her position. She may be sinking and only have time to send one message. If you are the only vessel to hear that message it is vital that you have a reliable written record of the essential details.

Acknowledgement of Receipt of a Distress Message
All communications concerned with rendering immediate assistance to the vessel in distress are prefixed with the distress signal 'MAYDAY'.
The acknowledgement of receipt of a distress message is given in the folalowing form —
— the distress signal 'MAYDAY';
— the callsign of the station sendinng the distress message, 3 times;
— the prowords 'THIS IS';
— the callsign of the station acknowledging receipt, 3 times;
— the word 'RECEIVED';
— the distress signal 'MAYDAY';
Example: —

MAYDAY – YACHT CATSPAW – CATSPAW – CATSPAW – THIS IS YACHT ECLIPSE – ECLIPSE – ECLIPSE – RECEIVED – MAYDAY –

Follow-up Information Message
As soon as a ship station has acknowledged receipt of a distress message, she must transmit as soon as practicable the following information in the order shown —
— the distress signal 'MAYDAY';
— the callsign of the station in distress, 3 times;
— the prowords 'THIS IS';

44

— the callsign of the calling station, 3 times;
— the position of the calling station;
— the speed at which she is proceeding to the station in distress;
— the approximate time it will take to reach the station in distress

Example:—

MAYDAY – CATSPAW – CATSPAW – CATSPAW – THIS IS ECLIPSE – ECLIPSE – ECLIPSE – MY POSITION ONE EIGHT ZERO FROM ST GOVANS HEAD ONE MILE – SPEED ONE FIVE KNOTS – WILL REACH YOU AT APPROXIMATELY ZERO THREE TWO FIVE – OVER.
Catspaw, if able could reply:—

MAYDAY – ECLIPSE – THIS IS – YACHT CATSPAW – MESSAGE UNDERSTOOD – WILL FIRE FLARES AT INTERVALS – OVER.
Eclipse should acknowledge:—

MAYDAY – CATSPAW – THIS IS ECLIPSE – UNDERSTOOD – OUT.

Mayday Relay procedure
A ship station or a Coast Radio Station which learns that a vessel is in distress, must transmit a distress message in any of the following cases:

1. when the station in distress cannot itself transmit a distress message;
2. when it is considered that further help is necessary;
3. when, although not in a position to render assistance, she has heard a distress message which has not been acknowledged.

When a distress message is being transmitted by a station, not herself in distress, this fact must be made quite clear. If this is not done, direction-finding bearings might be taken on the station making this transmission and assistance could thereby be directed to the wrong position. Therefore, when the station sending the distress message is not actually in distress itself, and in any other circumstances where a distress message might be repeated by a station not itself in distress, the transmission of the distress message must always be preceded by the following call:
— the signal 'MAYDAY RELAY', spoken three times;
— the prowords 'THIS IS' —
— the callsign or other identification of the station making

the transmission, spoken three times. For example let us imagine that the ship *Endurance* heard *Catspaw* transmitting her distress message. *Endurance* is not in a position to render assistance herself **and has heard no other acknowledgement** of *Catspaw's* distress message. *Endurance* decides to retransmit the distress message.

Endurance transmits: —

MAYDAY RELAY – MAYDAY RELAY – MAYDAY RELAY – THIS IS ENDURANCE – ENDURANCE – ENDURANCE – MAYDAY – YACHT CATSPAW ONE EIGHT ZERO FROM CALDY ISLAND ONE MILE – STRUCK ROCK IS DISABLED AND IN SINKING CONDITION – REQUIRES IMMEDIATE ASSISTANCE – CREW OF FOUR STILL ON BOARD BLUE MOTOR YACHT DRIFTING SOUTHWEST – OVER.

A yacht should not acknowledge receipt of a distress message transmitted by a Coast Radio Station as a MAYDAY RELAY under the conditions mentioned above unless she is in a position to render assistance.

Control of Communications During Distress

The responsibility for the control of distress traffic, which includes all communications concerned with rendering immediate assistance to a vessel in distress, lies with —

either, the station in distress;

or, the station sending the distress message on behalf of a ship in distress, i.e. MAYDAY RELAY;

or, A Coast Radio Station or Coastguard station which has responsibility delegated to it by the station in distress.

In coastal waters it is preferable to delegate the responsibility to the Coastguard or a Coast Radio Station. They have better facilities available to control the frequency being used in distress, and the Coastguard has direct communication with the Search and Rescue organisations.

Imposing Radio Silence

The station controlling distress traffic may impose silence either on 'All Stations' or any individual station which interferes with distress traffic. To impose silence it transmits: —

MAYDAY – SEELONCE MAYDAY – SEELONCE MAYDAY – SEELONCE MAYDAY – THIS IS SEVERN RADIO – SEVERN RADIO – OUT.

The expression 'SEELONCE MAYDAY' is reserved for the use of the station controlling distress traffic and no other station may use this expression.

If any other station near to the station in distress believes it essential to do so, it may impose silence, but in this case it must use the expression 'SEELONCE DISTRESS'.

To impose silence the station transmits:—

MAYDAY – SEELONCE DISTRESS – SEELONCE DISTRESS – SEELONCE DISTRESS – THIS IS ECLIPSE – ECLIPSE – OUT.

Note the difference between the two expressions used to impose radio silence.

SEELONCE MAYDAY — Station **controlling** distress traffic imposing silence.

SEELONCE DISTRESS — Station near to the station in distress, believing it essential to do so, imposing silence.

All stations which are aware of distress traffic, and are not taking part in it, are forbidden to transmit on the frequency or channel being used for distress except in the circumstances described below.

Relaxing Radio Silence

When distress traffic is being handled on Channel 16 all normal communication is suspended. The distress frequency is also the International calling frequency and, while the distress incident is being handled, delays in handling normal traffic are inevitable.

When complete silence is no longer considered necessary, the station controlling distress traffic will indicate that restricted working may be resumed by making the following transmission on the distress frequency:

— the distress signal 'MAYDAY';
— the call 'HELLO ALL STATIONS', spoken 3 times;
— the prowords 'THIS IS';
— the call sign or other identification of the station sending the message;
— the time of the message;
— the name and callsign of the mobile station which is in distress;
— the French word 'prudence' (PRU-DONCE)

Example:—

MAYDAY – HELLO ALL STATIONS – HELLO ALL STATIONS – THIS IS
SEVERN RADIO – SEVERN RADIO – TIME ZERO THREE FOUR FOUR –
YACHT CATSPAW – PRU-DONCE – OUT.

Cancelling Radio Silence

When the distress traffic has completely ceased, the station
which has controlled the distress traffic must let all stations
know that working may be resumed. This is done by sending
a message in the following form to 'ALL STATIONS'.
— the distress signal 'MAYDAY';
— the call 'HELLO ALL STATIONS', spoken 3 times;
— the prowords 'THIS IS';
— the callsign or other identification of the station sending
the message;
— the time of the message;
— the name and callsign of the mobile station which was in
distress;
— the words 'SEELONCE FEENEE'. (Again derived from the
French language)

MAYDAY – HELLO ALL STATIONS – HELLO ALL STATIONS – HELLO
ALL STATIONS – THIS IS SEVERN RADIO – SEVERN RADIO – TIME
ZERO FOUR FIVE ZERO – YACHT CATSPAW SEELONCE FEENEE –

Direction Finding

One additional type of signal may be used during distress
working. Lifeboats and some SAR aircraft are fitted with
Direction Finding receivers and may request a yacht in
distress to transmit a signal suitable for direction finding.
For example, a lifeboat going to the assistance of a yacht in
distress and wishing to take a D/F bearing would transmit —

MAYDAY – YACHT CATSPAW – CATSPAW – CATSPAW – THIS IS
MUMBLES LIFEBOAT – MUMBLES LIFEBOAT – MUMBLES LIFEBOAT
– FOR D/F PURPOSES WILL YOU HOLD YOUR 'PRESS TO SPEAK'
SWITCH CLOSED FOR TWO PERIODS OF TEN SECONDS EACH –
FOLLOWED BY YOUR CALLSIGN – AND REPEAT FOUR TIMES ON
THIS FREQUENCY – OVER.

The reply to this request should be —

MAYDAY – MUMBLES LIFEBOAT – MUMBLES LIFEBOAT – MUMBLES LIFEBOAT – THIS IS – YACHT CATSPAW – CATSPAW – CATSPAW –
(10 sec – 10 sec) CATSPAW –
(10 sec – 10 sec) CATSPAW –
(10 sec – 10 sec) CATSPAW –
(10 sec – 10 sec) CATSPAW –
OVER.

The request for a transmission for D/F may be repeated at intervals as the lifeboat closes the yacht.

CHAPTER VII

URGENCY AND SAFETY,
LIAISON WITH H.M. COASTGUARD ETC.

Where there is no IMMINENT danger to a ship or person and IMMEDIATE assistance is not required or fully justified then use can be made of the Urgency Signal (PAN PAN). This signal has priority over all other communications except distress.

The Urgency signal indicates that the station sending it has a **"very urgent message to transmit concerning the safety of a ship, aircraft (or other vehicle) or person."**

The Urgency Signal, Urgency Call and the message which follows are sent on 2182 kHz (MF) or on the VHF Channel 16. Note that any other frequency may be used in cases such as Distress and Urgency.

The Urgency call and Urgency message consists of the following —

— the Urgency Signal 'PAN PAN' spoken 3 times;
— the call 'HELLO ALL STATIONS', spoken up to 3 times;
— the prowords 'THIS IS'
— the callsign or other identification of the station sending the message, spoken up to 3 times;
— its position;
— the nature of the urgency;
— assistance required;
— the invitation to reply and acknowledge.

Example:—

PAN PAN – PAN PAN – PAN PAN – HELLO ALL STATIONS – HELLO ALL STATIONS – HELLO ALL STATIONS – THIS IS YACHT JASMINE – YACHT JASMINE – TWO FOUR ZERO FROM PORTLAND BILL LIGHT ONE FIVE MILES – COMPLETE ENGINE FAILURE – DRIFTING EAST AT THREE KNOTS – REQUIRE TOW URGENTLY – OVER

That is an example of a situation where the safety of a ship is at stake but a **MAYDAY** signal is not justified because the ship is not in imminent danger and immediate assistance is not required.

In the example above the urgency call is addressed to all stations. It may, however, be addressed to a single station if the vessel sending the signal knows which station is most likely to be able to provide the necessary assistance.

Medical Emergency
Where an Urgency Call involves a request for medical advice the Urgency Signal and Urgency Call should be sent on Channel 16 (Distress, Safety and Calling) and should announce the intention to switch to a working frequency in order to transmit the long message which is likely to be associated with obtaining medical advice. The point of this is that as soon as a United Kingdom Coast Radio Station is aware that you need medical advice the personnel there will be taking immediate steps to link you with a Doctor or Hospital direct. You save time then by transmitting as follows: —

PAN PAN MEDICO – PAN PAN MEDICO – PAN PAN MEDICO – ANGLESEY RADIO – ANGLESEY RADIO – THIS IS YACHT JASMINE – YACHT JASMINE – HAVE LONG URGENCY MESSAGE FOR YOU – CHANNELS TWENTY SIX OR TWENTY EIGHT – OVER .

Anglesey Radio would then inform you which Channel is free to use as the working frequency.
Medical advice can be obtained from any Coast Radio Station in the UK and Irish Republic (but note that not all Irish Coast Radio Stations are fitted with VHF) by a request to the station concerned. The Coast Radio Station will arrange a radiotelephone link with a hospital. Similarly, if medical assistance (e.g. a doctor) from shore is required, the request should be addressed as an Urgency Signal to the Coast Radio Station, which will pass the message to the Coastguard who will take the necessary action.
In both cases the messages will be exchanged free of charge, and the use of the Urgency Signal 'PAN-PAN' is proper in both cases.
The procedure for other countries is briefly: —

Belgium
Ships should call Oostende coast radio station using the address "Radiomédical Oostende". English, French, Dutch or German languages may be used.

Federal Republic of Germany
Ships should call the nearest coast radio station using the address "Funkarzt . . . (name of coast radio station)". English or German languages may be used.

France
Ships should call the nearest coast radio station using the address "PAN PAN (repeated three times) Radiomédical . . . (name of coast radio station)." French language should be used.

Netherlands
Ships should call Scheveningen coast radio station using the address "Radiomédical Scheveningen." English, Dutch, French or German languages may be used.

Republic of Ireland
As for United Kingdom.

It should be noted that Urgency messages may be addressed to a particular station or to All Stations. If you address your message to 'ALL STATIONS' then you must cancel it by a similarly addressed message when action is no longer necessary.

The above procedure is also necessary if a MAYDAY is being handled on 2182 kHz MF or Channel 16 VHF, and the need to transmit an Urgency Message arises. Listen carefully for a break in the distress traffic, make the Urgency Signal and Urgency Call and announce the intention to change to a working frequency to transmit the Urgency message.

The Safety Signal
The radiotelephone Safety Signal consists of the word 'SÉCURITÉ' (pronounced 'SAY-CURE-E-TAY') sent three times before the call and indicates that the station is about to transmit a message containing an important navigational or meteorological warning.

The Safety Signal and Call should be sent on either or both of the international distress frequencies, 2182 kHz MF or Channel 16 VHF, but may be sent on any other frequency designated for distress.

The safety message itself is not sent on an international distress frequency, but on a working frequency which is designated in the safety call.

SÉCURITÉ – SÉCURITÉ – SÉCURITÉ – HELLO ALL STATIONS – HELLO ALL STATIONS – THIS IS – LANDS END RADIO – LANDS END RADIO – NAVIGATIONAL WARNING – SWITCH TO CHANNEL TWO SEVEN –

On Channel 27 —

SÉCURITÉ – SÉCURITÉ – SÉCURITÉ – ALL STATIONS – ALL
STATIONS – THIS IS – LANDS END RADIO – LANDS END RADIO –
BREAKSEA LIGHT VESSEL OFF STATION FROM ONE SIX THREE
ZERO HOURS ON TWO JUNE FOR ONE TWO HOURS – OUT.

Safety messages are normally addressed to 'ALL STATIONS'.
They may, however, be addressed to a particular station.
All stations hearing the Safety Signal must listen to the
safety message until they are satisfied that it is of no
concern to them. They must not make any transmission
likely to interfere with the message.
The Safety Signal (announcing a safety message to follow)
is normally transmitted to all stations by a Coast Radio
Station on 2182 kHz and Channel 16 VHF, immediately on
receipt. It is then repeated at a time designated for the
broadcast of navigational warnings. These times differ
between stations but details are given in Yachtsmen's
Nautical Almanacs.

Meteorological Information
Particulars of stations sending out meteorological bulletins at
fixed times are given in the Admiralty List of Radio Signals
Vol. 3 and in Yachtsmens' Nautical Almanacs.
Met. warning messages are prefixed by the safety signal and
are normally transmitted on a working frequency, after a
preliminary announcement on the Distress, Safety and
Calling frequency.
A special Met. forecast for commercial shipping for any area
between parallels 35 degrees and 65 degrees North, and the
meridian 40 degrees West and the coasts of the European
Continent, for periods of up to 24 hours, may be obtained by
a commercial vessel through a UK Coast Radio Station. The
request is addressed to the Coast Radio Station and states
the required period, the required area and the ship's name.
The request will be sent to the Met. Office at Bracknell and
the reply sent to the ship by the coast Radio Station.
Radiotelephone link calls requesting forecasts are made
through UK Coast Stations to the Deputy Forecaster,
Bracknell 20242, Extension 2508. This service is not
available to pleasure yachts, but yachts may make link calls
to local weather centres to obtain forecasts.

Times and frequencies of weather forecasts and phone numbers of Weather Centres are available in RYA booklet G5.

Gale Warnings

Gale warnings are broadcast by UK Coast Radio Stations at the end of the first silence period (hour + 03 minutes and hour + 33 minutes) after receipt. They are then repeated at fixed times, 0303, 0903, 1503 and 2103, GMT for most stations but one or two have variations (i.e. for Jersey Radio the times are 0307, 0907, 1507 and 2107 GMT). The gale warning service on BBC Radio 4 is not available at night when Radio 4 has closed down and the times of broadcasts depend upon the scheduling of programmes. The gale warning service from Coast Radio Stations therefore offers advantages in terms of a 24-hour service; broadcasts at regular and predictable times and offers no more than half an hour's delay after receipt until the time of broadcast of a warning.

Navigational Warnings

Particulars of stations making regular transmissions of navigational information for the benefit of mariners are published in Yachtsmens' Nautical Almanacs.

HM Coastguard Maritime Rescue Centres

HM Coastguard has established Maritime Rescue Centres and Sub-Centres which are shown on H.M. Coastguard pamphlets. All these keep a constant watch on VHF Channel 16 and have the capability to operate on Channel 67 which is the Small Boat Safety frequency. It is of course obligatory, as we have said before, to have Channel 16 capability and you are strongly advised to ensure that your equipment is **also** capable of transmitting on Channel 67.

Small Craft Safety Channel 67

In the U.K. Channel 67 is reserved for small craft to speak, *on matters of safety,* direct to H.M. Coastguard. Considerable traffic may be heard at busy "yachting" times. Whilst H.M. Coastguard invite yachts on passage to notify them prior to departure under the Yacht and Boat Safety Report Scheme, they will also provide — as an aid to safety — a report of weather forecasts over the air on Channel 67. H.M. Coastguard issue Yacht and Boat Safety Scheme leaflets which explain the system of surveillance. All you need to do is fill in a simple postcard. You can get one from

Coastguard stations, marinas, yacht clubs, harbourmasters' offices and wherever you see the circular "Issuing Authority" sign.

Describe your craft, its equipment and your normal sailing area, and then drop the card into a post-box. Remember to up-date the information whenever details change.

In the ordinary course of casual coastal cruising you are not really recommended to clutter up the Channel with unnecessary correspondence on Channel 67, but it is worth passing details of longer passages.

Let us assume that you are on board your yacht *Audacity* intending to sail from Brighton Marina to Holyhead, North Wales. You set sail and, once clear of the port decide to notify H.M. Coastguard of your departure and also to check that your radiotelephone is working.

Listen on Channel 16 to make sure that your call will not interfere with other communications already in progress and then transmit:—

SHOREHAM COASTGUARD – THIS IS YACHT AUDACITY – YACHT AUDACITY – OVER.

H.M. Coastguard might reply as follows:—

AUDACITY – THIS IS SHOREHAM COASTGUARD – CHANNEL SIX SEVEN – OVER.

Having switched to Channel 67 and again listened for other communications in progress you might report your position and intentions:—

SHOREHAM COASTGUARD – THIS IS AUDACITY – YOU HOLD MY SAFETY SCHEME CARD – DEPARTED BRIGHTON – BOUND HOLYHEAD – OVER.

And Shoreham Coastguard would acknowledge.

Action Required on Arrival at Destination
When you arrive at your destination you must inform HM Coastguard having once alerted them to the fact that you are making passage. Once again you would transmit on Channel 16 having listened to make sure that you are not interfering with other traffic. You might, on this subsequent occasion call the Coastguard Rescue Centre nearest your point of arrival (i.e. Holyhead in our example) and your message, **ultimately** given on Channel 67, might read as follows:—

HOLYHEAD COASTGUARD – THIS IS YACHT AUDACITY – ARRIVED HOLYHEAD – CLOSING DOWN RADIO WATCH – OVER.

H.M. Coastguard would probably reply: —

AUDACITY – THIS IS HOLYHEAD COASTGUARD – YOUR SAFE ARRIVAL RECORDED – OUT.
Whilst this would be quite sufficient you could correctly finish by transmitting: —

HOLYHEAD COASTGUARD – THIS IS AUDACITY – OUT.
Whilst it may be very very tempting to pass short messages directly on Channel 16 don't forget that it is forbidden for even short messages except distress traffic hence the reason for transferring to Channel 67.

APPENDIX A
TABLE OF TRANSMITTING FREQUENCIES IN THE 156—174 MHz BAND FOR STATIONS IN THE MARITIME MOBILE SERVICE

Channel desig nators	Transmitting frequencies (MHz)		Inter ship	Port operations		Ship movement		Public corres pon dence
	Ship stations	Coast stations		Single fre quency	Two fre quency	Single fre quency	Two fre quency	
60	156.025	160.625			•		•	•
01	156.050	160.650			•		•	•
61	156.075	160.675			•		•	•
02	156.100	160.700			•		•	•
62	156.125	160.725			•		•	•
03	156.150	160.750			•		•	•
63	156.175	160.775			•		•	•
04	156.200	160.800			•		•	•
64	156.225	160.825			•		•	•
05	156.250	160.850			•		•	•
65	156.275	160.875			•		•	•
06	156.300		•					
66	156.325	160.925			•		•	•
07	156.350	160.950			•		•	•
67	156.375	156.375	•	•		•		
08	156.400		•					
68	156.425	156.425		•		•		
09	156.450	156.450	•	•		•		
69	156.475	156.475	•	•		•		
10	156.500	156.500	•	•		•		
70	156.525		DIGITAL SELECTIVE CALLING					

Channel designators	Transmitting frequencies (MHz)		Inter ship	Port operations		Ship movement		Public correspondence
	Ship stations	Coast stations		Single frequency	Two frequency	Single frequency	Two frequency	
11	156.550	156.550		•		•		
71	156.575	156.575		•		•		
12	156.600	156.600		•		•		
72	156.625		•					
13	156.650	156.650	•	•		•		
73	156.675	156.675	•	•		•		
14	156.700	156.700		•		•		
74	156.725	156.725		•		•		
15	156.750	156.750	•	•				
75			Guard-band 156.7625 – 156.7875 MHz					
16	156.800	156.800	**DISTRESS Safety and Calling**					
76			Guard band 156.8125 – 156.8375 MHz					
17	156.850	156.850	•	•				
77	156.875		•					
18	156.900	161.500			•		•	
78	156.925	161.525			•		•	•
19	156.950	161.550			•		•	
79	156.975	161.575			•		•	
20	157.000	161.600			•		•	
80	157.025	161.625			•		•	
21	157.050	161.650			•		•	
81	157.075	161.675			•		•	•
22	157.100	161.700			•		•	
82	157.125	161.725			•		•	•
23	157.150	161.750						•
83	157.175	156.175 or 161.775						•
24	157.200	161.800						•
84	157.225	161.825			•		•	•
25	157.250	161.850						•
85	157.275	161.875						•
26	157.300	161.900						•
86	157.325	161.925						•
27	157.350	161.950						•
87	157.375	161.975						•
28	157.400	162.000						•
88	157.425	162.025						•

In the United States of America, Channels 07 18 19 21 22 23 65 66 78 79 80 83 and 88 operate in the simplex mode, using the ship station transmitting frequencies as follows:

Channel designator	Transmitting frequencies (MHz)	
	Ship stations	Coast stations
07A	156.350	156.350
18A	156.900	156.900
19A	156.950	156.950
21CG	157.050	157.050
22CG	157.100	157.100
23CG	157.150	157.150
65A	156.275	156.275
66A	156.325	156.325
78A	156.925	156.925
79A	156.975	156.975
80A	157.025	157.025
83CG	157.175	157.175
88A	157.425	—

NOTES

1. Intership Frequencies, Channels 06, 08, 72 and 77.
When selecting intership frequencies those allocated solely for intership use, i.e. Channels 06, 08, 72 and 77, should be chosen in preference to frequencies which are also designated for use for other purposes. However, please see also note 3 on other uses of channel 06.

2. Search and Rescue Co-ordination and Anti-pollution Operations, Channels 06, 67 and 73
Channel 06 may be used for communication between ships and aircraft engaged in co-ordinated search and rescue operations. Intership traffic must not be exchanged on channel 06 if interference with search and rescue co-ordination might result.
Channels 10, 67 and 73 may also be used for communication between ships, aircraft and land stations for search and rescue co-ordination and for anti-pollution operations.

3. Small Craft Safety, Channel 67

Within the U.K. channel 67 may be used for communication between small craft and the Coastguard on matters relating to safety.

4. Radio Lighthouses

There are two trial radio lighthouses now in operation. The VHF transmissions provide the bearing of the lighthouse as follows. A VHF signal is modulated with a tone which can be heard on a VHF receiver tuned to Channel 88. Along one particular bearing no tone is heard (the null radial), and the radial at North Foreland rotates slowly over the sector 180° to 300° (bearings to the lighthouse from seaward). After an initial long dash the tone is broken up into 67 beats, and each beat corresponds to a radial at intervals of 2°. The bearing to the Lighthouse is obtained by counting the beats until the tone disappears. The volume of the tone decreases as the null radial is approached and increases again following the null.

The details of this service are subject to change. The use of VHF frequencies is only temporary.

APPENDIX B
Deleted in 1991 Reprint.

APPENDIX C
Examination Centres for the Restricted Certificate of Competence (VHF only)

Southern
Southampton Institute of Higher Education,
Warsash Campus, Newtown Road,
Southampton, Hants.
Tel: 04895 76161

Isle of Wight
Club UK, Arctic Road, Cowes,
Isle of Wight.
Tel: 0983 294941

South West
Institute of Marine Studies, Polytechnic South West,
Drake Circus, Plymouth PL4 8AA.
Tel: 0752 226042

Falmouth Technical College,
Killigrew Street, Falmouth, Cornwall.
Tel: 0326 313326

Dept ACE, Brunel Technical College,
Ashley Down, Bristol, Avon.
Tel: 0272 241241 Ext. 2190

London
London Guildhall University, Department of Business Studies
84 Moorgate, London EC2M 6SQ
Tel: 071-320 1777 or 1451

East Anglia
West Mersea Yacht Club, 116 Coast Road,
West Mersea, Colchester CO5 8PB.
Tel: 0206 382947

Wales
Glamorgan Institute of Higher Education,
Department of Maritime Studies,
Western Avenue, Cardiff CF5 2YB.
Tel: 0222 551111

Yorkshire and Humberside
School of Engineering,
Humberside College of Higher Education,
Queens Gardens, Hull, HU1 3DH.
Tel: 0482 224121 Ext. 283/200

Tyneside
South Tyneside College,
Department of Nautical Science, St. George's Avenue,
South Shields, Tyne & Wear. NE34 6ET
Tel: 0632 560403

Scotland
RYA Scotland, Caledonia House, South Gyle,
Edinburgh EH12 9DQ
Tel: 031 317 7388

North West
Riversdale College of Technology,
Navigation Department, Riversdale Road,
Liverpool 19 3QR.
Tel: 051 427 1227

Northern Ireland
School of Maritime Studies, Ulster Polytechnic,
Shore Road, Newtonabbey, Co. Antrim BT37 0QB.
Tel: 0231 65131

Jersey
VHF Radio Examinations, Highlands College, P.O. Box 142,
Jersey, C.I.
Tel: 0534 71065 Ex. 354

Guernsey
College of Further Education, Route de Coutanchez, St. Peter Port,
Guernsey, C.I.
Tel: 0481 27121

For Serving Servicemen
Examinations are administered by service Sailing Associations.

APPENDIX D
Radio Publications of interest to Yachtsmen

RYA Booklet G26 VHF Radio Operator Examinations
Gives the syllabus, likely questions and examination arrangements.

Yachtsmens' Nautical Almanacs
Reed's, Macmillan & Silk Cut and the Channel West & Solent Nautical Almanacs all contain sections on radio communications, including frequencies of Coast Radio Stations and Port Operations for individual harbours.
It is therefore possible to obtain the necessary radio information without recourse to specialist publications.

The Handbook for Radio Operators
This handbook is intended for the guidance of radio operators in ship and coast stations operating radio equipment on frequencies in the international maritime mobile bands and is subject to the overriding authority given in the International and National Regulations detailed below.
It is made up of the following eight chapters:

Chapter I   General Regulations and Conditions to be Observed by Stations of the Maritime Mobile Service.

Chapter II   Radiotelegrams, Radiotelephone and Radiotelex Calls.

Chapter III   Selective Calling.

Chapter IV   Procedures in the Maritime Mobile Radiotelegraph Service.

Chapter V   Distress, Urgency and Safety Communications by Radiotelegraphy.

Chapter VI   Procedures in the Maritime Mobile Radiotelephone Service.

Chapter VII   Distress, Urgency and Safety Communications by Radiotelephony.

Chapter VIII   Special Services.

The handbook is based on the provisions of:
(a) the International Telecommunication Covention (Malaga-Torremolinos, 1973);
(b) the Radio Regulations and Additional Radio Regulations, Geneva, 1967, as revised by the World Administrative

Radio Conference to deal with matters relating to the maritime mobile service, Geneva, 1974;

(c) the Merchant Shipping (Radio) Rules, 1965;
(d) the Merchant Shipping (Direction-Finders) Rules, 1965;
(e) the Merchant Shipping (Radio) (Fishing Vessels) Rules, 1974;
(f) the Wireless Telegraphy Acts, 1949 to 1967;
(g) the Telegraph Regulations and Telephone Regulations (Geneva 1973);
(h) on certain other statutory provisions.

Admiralty List of Radio Signals (ALRS), Vols 1-6
ALRS are the world-wide authoritative source of radio information for mariners. They are kept up to date by the weekly editions of Admiralty Notices to Mariners.
There is little point in carrying these books in a yacht cruising only in the waters around the British Isles but for cruising in distant waters certain volumes are almost essential.
The information contained in each volume is: —

Vol. I
Coast Radio Stations; Medical Advice by Radio; arrangements for Quarantine Reports and Locust Reports; Regulations for the use of Radio in Territorial waters; Distress, Search and Rescue procedure; the AMVER Organization and a brief extract from the International Radio Regulations.

Part 1 covers Europe, Africa and Asia (excluding the Philippine Islands and Indonesia)
Part 2 covers the Philippine Islands, Indonesia, Australasia, the Americas, Greenland and Iceland.

Vol. II
Radiobeacons (including Aero Radiobeacons in coastal regions); Radio Direction-finding Stations; Coast Stations which give a QTG service. Calibration Stations (i.e. stations giving special transmissions for the calibration of ships' DF); and Radar Beacons (Racons and Ramarks). Volume 2a contains disgrams only, for use in conjunction with Volume 2.

Vol III
Radio Weather Services and related information, including certain Meteorological Codes provided for the use of shipping.
Volume IIIa contains diagrams only, for use in conjunction with Volumes III and IV

Vol. IV
List of Meteorological Observation Stations.

Vol. V
Standard Times; Radio Time Signals; Radio Navigational Warnings (including relevant Codes and Practices, and Ice Reports); and Electronic Position-fixing Systems (Loran, Consol, Decca, Omega and Satellite Navigation). Volume 5a contains diagrams only, for use in conjunction with Volumes 5 and 6.

Vol. VI
Stations working in the Port Operations and Information Services; services to assist vessels requiring Pilots; and services concerned with Traffic Management.
Part 1 covers NW Europe and the Mediterranean
Part 2 covers Africa and Asia (excluding Mediterranean Coasts,) Australasia, Americas, Greenland and Iceland.

Standard Marine Navigational Vocabulary (Merchant Shipping Notice M 1252).
A vocabulary, compiled to standardise the language used in communication for navigation at sea, in port-approaches, in waterways and habours, compiled by the Inter-Governmental Maritime Consultative Organisation.

APPENDIX E
A SPECIMEN PROCEDURE CARD

YACHT'S NAME —
CALLSIGN —

TO SEND A DISTRESS SIGNAL

1. CHECK MAIN YACHT'S BATTERY SWITCH "ON"
2. SWITCH ON SET and turn power switch to "HIGH"
3. SWITCH TO CHANNEL 16
4. PRESS MICROPHONES SWITCH IN ORDER TO SPEAK
5. TRANSMIT SLOWLY AND CLEARLY:—

"MAYDAY MAYDAY MAYDAY"

THIS IS —
YACHT.....YACHT.....YACHT..... (3 times)

MAYDAY — YACHT.....

MY POSITION IS

(then give position as latitude and longitude or bearing **FROM** and range **OF** a conspicuous charted feature)

Then describe nature of distress:—

e.g. AM SINKING

Then give any other information which may help rescuers.

Then say "OVER" and **RELEASE** the "Press to Speak" switch.

IF YOU GET NO IMMEDIATE REPLY, CHECK RADIO AND TRY AGAIN.

NOTES

NOTES